THIS NOTEBOOK BELONGS TO :

....

....

YEAR :

....

Made in the USA
Las Vegas, NV
07 October 2021

31923025R00070